FLORA OF TROPICAL EAST AFRICA

MALPIGHIACEAE

E. Launert
(British Museum (Natural History))

Mostly woody climbers, sometimes shrubs or small trees, with unicellular appressed (sometimes fork-shaped) medifixed ± stiff hairs. Leaves opposite, ternate or alternate, simple and entire, often with glands near the base of the lamina or on the petiole; stipules present or absent. Inflorescence terminal or axillary, usually many-flowered and racemose (or more rarely flowers solitary); bracts and bracteoles present. Flowers actinomorphic or zygomorphic, bisexual (in African genera). Sepals 5, free or connate at the base, persistent, often with glands outside. Petals 5, frequently clawed, free, imbricate, entire or with fringed or dentate margins. Stamens 10, with those of the outer series opposite the petals; filaments often connate at the base; anthers dehiscing longitudinally, introrse, basifixed or dorsifixed, 2-thecous. Ovary superior, syncarpous, 3(rarely 2, 4, or 5)-locular and -lobed, with 1 pendulous axile ovule in each locule; styles as many as the carpels with usually entire stigmas. Fruit a schizocarp, usually winged (samara), rarely a fleshy drupe. Seeds with a large usually straight embryo, without endosperm.

A large family of about 60 genera and 800 species, with a mainly pantropical distribution, but extending into the subtropics, most abundant in the New World.

Thryallis glauca (Cavan.). O. Ktze. (*Galphimia glauca* Cavan.), an ornamental shrub with rather small glabrous elliptic pointed opposite leaves, bright yellow petals and red filaments, and small unwinged 3-locular fruits, is cultivated in Kenya (e.g. Nairobi Arboretum, *G. R. Williams* 453), Tanganyika (e.g. Amani, *G. R. Williams* 654) and Zanzibar (U.O.P.Z.: 270 (1949)).

Leaves spirally arranged:
  Calyx with 1 to several glands outside; mericarps
    with a straight or oblique ± parallel-nerved
    dorsal wing; lateral wing entirely reduced   .   1. **Acridocarpus**
  Calyx without glands; mericarps with a shield-like
    circular lateral wing; dorsal wing much reduced,
    crest-like .   .   .   .   6. **Caucanthus**
Leaves opposite:
  Anthers linear, 3–4 mm. long; mericarps with lateral
    wing divided into 5–7 narrow stellately arranged
    lobes .   .   .   .   2. **Tristellateia**
  Anthers ovate, oblong or ovate-oblong, 1–1·8
    (–2·5) mm. long; mericarps with wings entire or
    only emarginate at apex:
    Ovary glabrous .   .   .   .   .   3. **Triaspis**
    Ovary sericeous or tomentose:
      Styles shorter than or as long as ovary and
        shorter than, as long as or only slightly
        exceeding stamens .   .   .   .   6. **Caucanthus**
      Styles always longer than ovary and distinctly
        exceeding stamens:

Sepals not enclosing petals in bud; petals
    clawed .   .   .   .   .   .   **3. Triaspis**
Sepals completely enclosing petals in bud;
    petals sessile or clawed:
    Stigmas terminal; anthers usually some-
        what pilose; petals clawed; sepals
        pubescent inside; samara with only
        1 mericarp developed, this sub-
        globose, drupe-like, with 3 parallel
        wings .   .   .   .   .   **4. Flabellariopsis**
    Stigmas lateral; anthers glabrous; petals
        not clawed; sepals glabrous inside;
        samara with 3 (rarely only 2) meri-
        carps developed, these with a shield-
        like lateral wing .   .   .   .   **5. Flabellaria**

1. ACRIDOCARPUS

Guill. & Perr. in Fl. Seneg. Tent.: 123, t. 29 (Sept. 1831),
nom. conserv. propos.

Heteropteris Kunth sect. *Anomalopteris* DC., Prodr. 1: 592 (1824)
Anomalopteris (DC.) G. Don, Gen. Syst. 1: 647 (Aug. 1831);
O. Ktze., Rev. Gen. Pl. 1: 87 (1891)

Erect, suberect, trailing or climbing shrubs, rarely small trees. Leaves
alternate, petiolate, entire, usually with glands on undersurface at base, and
sometimes with 2 rows of smaller glands parallel to the margins, exstipulate.
Inflorescences few- to many-flowered corymbs, racemes or panicles, axillary
or terminating leafy branches; bracts present and persistent, small; bracteoles at base of pedicels, sometimes with a circular gland at base. Flowers
actinomorphic or nearly so. Calyx ± coriaceous, with 1 or more subcircular
sessile or sunken glands; lobes 5, equal or subequal, obtuse. Petals 5, white
or yellow, usually clawed, longer than sepals, entire, fimbriate or lacerate.
Stamens 10; anthers basifixed, glabrous; filaments usually thick, somewhat
broadened and connate at base, glabrous. Ovary 3-locular, but usually with
1 locule abortive, usually densely sericeous or tomentose-sericeous; styles 2,
terete, curved inwards. Samara with a straight or oblique dorsal wing.

A genus of about 30 species, mainly occurring in tropical Africa; one species in
Madagascar, one in New Caledonia.

Bracteoles with a circular or elliptic gland at base,
    or glands on rhachis and attached to
    bracteoles (see fig. 2/7):
Bracts triangular, usually acute, as broad as
    or broader than long, 0·8–1 mm. long;
    rhachis of inflorescence robust, ± thick-
    ened (sausage-like) .   .   .   .   1. *A. alopecurus*
Bracts ovate, often obtuse-acuminate, longer
    than broad, up to 2·25 mm. long; rhachis
    of inflorescence not thickened:
Wing of samara obliquely obovate (see fig.
    1/9); inflorescences mostly terminal,
    stiff; pedicels ± robust; glands trans-
    versely elliptic or subcircular, affixed on
    rhachis below bracteoles (see fig. 2/7) .   2. *A. smeathmanii*
Wing of samara obliquely oblong to obovate-
    oblong (see fig. 1/3); inflorescences

mostly axillary, rarely terminal, very
loose; pedicels rather slender; bracteoles with a circular gland at the very
base . . . . .   3. *A. prasinus*
Bracteoles and rhachis eglandular:
 Leaf-lamina 0·5–1·25(–1·5) cm. broad and not
 longer than 6 cm., densely ferrugineous-
 villous or subtomentose or greyish-pubescent (mainly beneath); wing of samara up
 to 2(–2·5) cm. long . . . .   4. *A. glaucescens*
 Leaf-lamina much broader and always longer,
 glabrous beneath or rarely sparsely hairy
 or sericeous (often on or beside the midrib
 beneath), only very young ones sometimes
 ferrugineous-sericeous; wing of samara
 always more than 3 cm. long:
 Flowers in terminal corymbs (see fig. 2/1);
  bracts 3–3·25 mm. long . . .   9. *A. congestus*
 Flowers in terminal or axillary pyramidal
  usually elongate racemes; bracts 1–2·5
  (rarely –3) mm. long:
  Racemes terminating leafy branches, with
   rhachis (8–)10–30 cm. long:
   Leaf-lamina often conduplicate, base
    usually cuneate, very rarely
    rounded, quite glabrous, pale
    green when dried; petiole (0·6–)
    0·7–1·2 cm. long, glabrous . .   5. *A. zanzibaricus*
   Leaf-lamina not conduplicate, base
    rounded or sometimes subcordate,
    very rarely cuneate, ferrugineous-
    tomentose beneath at first, later
    only on or beside midrib, brown or
    dark green when dried; petiole 0·2–
    0·4(–0·6) cm. long, usually densely
    sericeous . . . . .   6. *A. chloropterus*
  Racemes mostly axillary, with rhachis
   1–8(–10) cm. long:
   Leaf-lamina 8–24 × 3·3–6·5(–8) cm.;
    wing of samara obliquely ovate to
    ovate-oblong or oblong (see fig.1/10)   7. *A. scheffleri*
   Leaf-lamina 4–7(–9) × 1·5–2·5(–3) cm.;
    wing of samara obliquely semi-
    obovate (see fig. 1/1) . .   8. *A. pauciglandulosus*

1. **A. alopecurus** *Sprague* in K.B. 1909: 185 (1909); Niedenzu in Arb. Bot.
Inst. Akad. Braunsberg 7: 14 (1921) & in E.P. IV. 141: 272 (1928);
T.T.C.L.: 295 (1949); Wilczek in F.C.B. 7: 234 (1958). Type: Pemba I.,
Barrand (K, holo.!)

Spreading shrub or woody climber. Young shoots ferrugineously sericeous,
older ones somewhat pubescent but usually glabrescent, lenticellate. Leaf-
lamina obovate, obovate-oblong, elliptic or oblanceolate, 11–18(–20) ×
2·8–6·5(–8·5) cm., shortly acuminate, rounded or cuneate at base, chartaceous
or chartaceous-membranous, usually glabrous but sometimes finely pubescent, with 7–11 lateral nerves, mostly with 1 or 2 glands beneath near
insertion of petiole, and often with a pair of smaller ones near the apex at

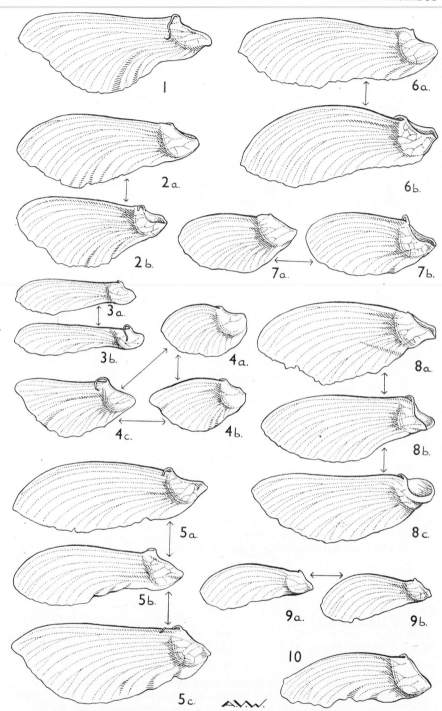

FIG. 1. Mericarps of various species of *ACRIDOCARPUS*, all × 1—**1**, *A. pauciglandulosus*; **2**, *A. chloropterus*; **3**, *A. prasinus*; **4**, *A. glaucescens* var. *ferrugineus*; **5**, *A. alopecurus* var. *alopecurus*; **6**, *A. congestus*; **7**, *A. ugandensis*; **8**, *A. zanzibaricus*; **9**, *A. smeathmanii*; **10**, *A. scheffleri*. 1, from *Schlieben* 5498; 2, from *Faulkner* 109; 3a, b, from *Bagshawe* 772; 4a, b, from *J. Adamson* 96; 4c, from *Puccioni & Stefanini* 246 (274); 5a, from *Paulo* 198; 5b, from *Braun in Herb. Amani* 3352; 5c, from *Semsei* 1431; 6, from *Eggeling* 6788; 7, from *Dale* K294; 8a, from *Napper* 463; 8b, from *Faulkner* 2394; 8c, from *Battiscombe* 801; 9, from *Procter* 707; 10, from *Greenway* 2945.

base of apiculum; petiole 4–8 mm. long, robust, canaliculate, finely pubescent or glabrous. Inflorescences mostly axillary (rarely terminal) racemes, many-flowered; axis very stout, often thickened (sausage-like), very short and very shortly peduncled or often elongate and then with a longer peduncle (up to 2 cm.), ± densely ferrugineous- or rubiginose-sericeous; bracts broadly ovate-triangular or lanceolate-triangular, usually acute, persistent, inserted at base of or some way below pedicel, with a pair of glands above or at the very base; bracteoles very small, sometimes with a single gland at base. Flowers ± 2·5 cm. in diameter. Calyx with usually 3 (rarely 5) circular glands; lobes subcircular or ovate to ovate-oblong, 3–3·5 mm. long, glabrous. Petals golden yellow, subcircular to broadly obovate, 10–12 × 7–9 mm., shortly clawed, deeply lacerate on the margin. Stamens with oblong anthers 3·5–5 mm. long; filaments cylindrical, 1–2 mm. long, glabrous. Ovary ± densely sericeous; styles 2, 0·7–1 cm. long. Wing of samara (see fig. 1/5) obliquely obovate to obovate-oblong, 3·5–4·5(–5·5) × 1·7–2·1 cm., with the upper margin straight or slightly curved towards the apex, very rarely somewhat constricted at the base, distally obliquely truncate, clasping the nut halfway or somewhat nearer to the base.

var. alopecurus

Inflorescences axillary with a short (1–2(–2·5) cm. long) axis, usually becoming sausage-like and thickened after flowering; bracts broadly ovate-triangular, up to 1 mm. long, usually acute, with a pair of distinct circular clearly visible glands (black in dried material) at the very base, inserted at base of pedicel. Annual shoots mostly soon glabrescent. Fig. 1/5.

KENYA. Kwale District: Shimba Hills, 14 Jan. 1964, *Verdcourt* 3920A!; Kenya coast, without precise locality, *R. M. Graham* in *F.D.* 1786!
TANGANYIKA. Pangani District: Pangani R. at Hale, 3 Dec. 1910, *Braun* in *Herb. Amani* 3352! & Bushiri Estate, 31 Dec. 1950, *Faulkner* 744!; Morogoro District: Turiani, Manyangu, Nov. 1953, *Paulo* 198! & Turiani, Nov. 1953, *Semsei* 1431!
ZANZIBAR. Pemba I., Weti, *Barrand*!
DISTR. **K**7; **T**3, 6; **P**; Congo Republic
HAB. Dry evergreen and riverine forest, particularly in more open parts and at margins; near sea-level to ± 150 m.

SYN. *A. alopecurus* Sprague subsp. *glaber* Niedenzu in Arb. Bot. Inst. Akad. Braunsberg 7: 14 (1921) & in E.P. IV. 141: 273 (1928); T.T.C.L.: 295 (1949). Types: Tanganyika, Pangani District, Pangani R. near Hale, *Braun* in *Herb. Amani* 1528 (B, syn. †, EA, K, isosyn.!) & Hale, *Braun* in *Herb. Amani* 3352 (B, syn. †, EA, isosyn.!) & Makinjumbe, *Scheffler* 245 (B, syn. †, EA, isosyn.!)
    A. alopecurus Sprague subsp. *pubescens* Niedenzu var. *typicus* in Arb. Bot. Inst. Akad. Braunsberg 7: 14 (1921) & in E.P. IV. 141: 273 (1928); T.T.C.L.: 295 (1949), *nom. illegit.* Type: as for species

   var. **machaeropterus** *Niedenzu* in Arb. Bot. Inst. Akad. Braunsberg 7: 14 (1921) & in E.P. IV. 141: 273 (1928), as a variety of subsp. *pubescens* Niedenzu; T.T.C.L.: 295 (1949). Type: Tanganyika, Uzaramo District, Pugu Hills, *Holtz* 2146 (B, holo. †); Pugu Forest Reserve, *Semsei* 1723 (EA, neo.!, K, isoneo.!)

Inflorescences terminating leafy branches, with an elongate (3–8 cm. long) very slightly thickened axis; bracts lanceolate-triangular, up to 2 mm. long, always acute, with a pair of usually very small glands mostly above the base, inserted some way below the pedicel. Annual shoots ± coriaceous, very tardily glabrescent.

TANGANYIKA. Uzaramo District: Pugu Hills, 12 Feb. 1938, *Vaughan* 2748! & Pugu Forest Reserve, June 1954, *Semsei* 1723! & Kazimzumbwi Forest Reserve, 5 Mar. 1964, *Semsei* 3655!
DISTR. **T**6 (known only from the Pugu Hills and nearby)
HAB. Little known, probably dry evergreen forest; 200–300 m.

2. A. smeathmanii (*DC.*) *Guill. & Perr.*, Fl. Seneg. Tent.: 124 (1831); A. Juss., Malpigh. Synops., in Ann. Sci. Nat., sér. 2, 13: 271 (1840) & in Archiv. Mus. Paris 3: 484, t. 15 (1843); Oliv., F.T.A. 1: 277 (1868), pro parte (var. *a*); Sprague in J.B. 1906: 204 (1906); Niedenzu in Arb. Bot. Inst. Akad. Braunsberg 6: 56 (1915) & 7: 17 (1921) & in Verz. Vorl. Akad. Braunsberg S.-Sem. 1924: 18 (1924) & in E.P. IV. 141: 274 (1928), excl.

syn. *A. chevalieri* Sprague; Wilczek in F.C.B. 7: 232 (1958); F.W.T.A., ed. 2, 1: 352, fig. 125 (1958). Type: Sierra Leone, *Smeathman* (P, holo., BM, iso. !)

Strong woody climber, up to 7 m. or more in length; older stems up to 12 cm. in diameter. Younger branches ± densely ferrugineous-sericeous or pubescent, older ones glabrous, lenticellate. Leaf-lamina obovate, obovate-oblong to obovate-elliptic, rarely oblanceolate, (4–)5·5–14(–17) × (2–)3–6(–9) cm., obtuse, abruptly acuminate, rarely somewhat retuse, cuneate or rounded at base, subcoriaceous, finely pubescent underneath (mainly along the midrib) when young, later glabrous, dark green and somewhat glossy on upper, pale green and matt on lower surface, with 8–12 pairs of lateral nerves and with 1 or 2 pairs of circular glands beneath near insertion of petiole; petiole 0·5–1 cm. long, canaliculate. Inflorescences many-flowered racemes which are usually aggregated into large panicles, 5–20(–30) cm. long; bracts triangular or ovate-triangular, 1–2·25 mm. long; bracteoles similar to the bracts, with a circular or transversely elliptic gland on rhachis below (fig. 2/7). Flowers ± 2·5 cm. in diameter. Calyx with 2 or 3 (rarely 4) circular glands at the very base; lobes broadly ovate to subcircular, 2·5–3 × 3 mm., finely sericeous outside or glabrescent. Petals yellow, obovate to elliptic, 9–14 × 7–10 mm., clawed, usually fimbriate at margins. Stamens with the anthers (3·5–)4–5 mm. long; filaments rather thick, somewhat dilated towards base, 1·5–3 mm. long. Ovary densely ferrugineous-sericeous; styles 8–10 mm. long. Wing of samara obliquely obovate-oblong, 2·5–5 × 0·9–1·5 cm., with upper edge curved, somewhat constricted near base, broadest above middle, green and tinged with purple. Figs. 1/9, p. 4, 2/7, p. 10.

TANGANYIKA. Buha District: [Kibondo–Kasulu road], Malagarasi R. ferry, Sept. 1957, *Procter* 707 !
DISTR. T4; Congo and Cameroun Republics west to Mali and Portuguese Guinea
HAB. Riverine forest; ± 1150 m.

SYN. *Heteropteris* (?) *smeathmanii* DC., Prodr. 1: 592 (1824)
    Anomalopteris spicata G. Don, Gen. Hist. 1: 647 (1831), *nom. illegit.* Type: as for species
    A. smeathmanii (DC.) O. Ktze. in Rev. Gen. Pl. 1: 87 (1891)
    Acridocarpus goossensii De Wild., Pl. Bequaert. 1: 233 (1922). Type: Congo Republic, Mayumbe, *Goossens* 1341 (BR, holo.)

NOTE. Due to its wide distribution this species shows a great range of variation which is most apparent in the shape and size of the leaves and the wing of the samara. The single East African gathering is a very poor specimen, but there is no doubt about its identity. The locality marks the furthest eastern point of the extensive distribution of *A. smeathmanii* yet known.

3. **A. prasinus** *Exell* in J.B. 65, Suppl. Polypet.: 51 (1927); Niedenzu in E.P. IV. 141: 278 (1928); Gossweiler & Mendonça in Cart. Fitogeogr. Angola: 145 (1939); Exell & Mendonça in C.F.A. 1: 253 (1951); Wilczek in F.C.B. 7: 232 (1958). Type: Angola, Zaire, Sumba, Peco, *Gossweiler* 9115 (BM, holo. !)

Woody climber or many-branched dense shrub, with spreading and scrambling branches. Younger branches ± densely rusty-tomentose but very soon glabrescent, lenticellate. Leaf-lamina oblong or oblong-elliptic, 3–12 × 1·75–4 cm., distinctly acuminate (sometimes ± caudate), cuneate, subcoriaceous, usually eglandular, ± densely (? reddish) sericeous when young, later usually quite glabrous except on the midrib beneath, with 6–8 pairs of lateral nerves; petiole rather slender, 0·5–1 cm. long, subcanaliculate, usually glabrous. Inflorescences terminal or axillary racemes, with a slender densely pubescent 1–2 cm. long rhachis, 9–12(–15)-flowered; bracts ovate-acuminate, up to 2 mm. long, pubescent, persistent; bracteoles ± 0·7 mm. long, with a circular gland at the base. Flowers ± 2·5 cm. in diameter,

Calyx with 2–3 ± distinct glands; lobes oblong-elliptic, 2–4 mm. long, brownish-sericeous outside or glabrescent. Petals white, subcircular or elliptic, 8–13 × 6–9 mm., shortly clawed. Stamens with oblong or oblong-ovate anthers 2·25–4 mm. long; filaments 0·7–1 mm. long, glabrous. Ovary brownish-tomentose; styles 2, slender, 5–11 mm. long. Wing of samara obliquely oblong-obovate (see fig. 1/3), 2·5–3 × 0·5–0·8 cm., usually with upper edge straight, broadest in the upper third, not clasping the nutlet. Fig. 1/3, p. 4.

UGANDA. Masaka District: Byante Central Forest Reserve, July 1951, *Philip* 483! ; Mengo District: near Entebbe, Kitubulu, Sept. 1945, *Eggeling* 5556 ! & 3 km. E. of Entebbe, Kyewaga Forest, 4 Nov. 1950, *F.D. staff* in *Dawkins* 672 !
DISTR. U4; Congo Republic, Angola
HAB. Margins and more open places in rain-forest; 1100–1200 m.

4. **A. glaucescens** *Engl.* in Ann. Ist. Bot. Roma 9: 253 (1902); Sprague in J.B. 44: 202 (1906); Niedenzu in Arb. Bot. Inst. Akad. Braunsberg 7: 5 (1921) & in E.P. IV. 141: 264 (1928); Chiov., Fl. Somala 1: 109 (1929) & 2: 44 (1932); E.P.A.: 404 (1956); Launert in K.B. 19: 351 (1965). Type: Somali Republic (S.), " Dar near Barden " [probably Bardera], *Riva* 200* (FI, holo. !)

Upright sparingly branched shrub, up to 2·5 m. high. Younger branches ± densely ferrugineous- or greyish-sericeous, older ones usually glabrous, lenticellate. Leaf-lamina linear, linear-lanceolate, oblanceolate, oblong-lanceolate to narrowly elliptic, 1·5–6 × 0·5–1·25(–1·5) cm., acute to obtuse, sometimes finely apiculate, cuneate or attenuate at the base, glabrous to densely hairy (see below), rigidly coriaceous, with revolute margins, with 8–12 pairs of lateral nerves, eglandular or rarely with a pair of small glands beneath near insertion of petiole; petiole 1–3 mm. long, canaliculate, glabrous or somewhat sericeous. Inflorescences on leafy branchlets terminating in few- to many-flowered pyramidal racemes, 3–8(–12) cm. long; rhachis robust, usually ± densely ferrugineous-sericeous; bracts subulate, 2–2·5 mm. long, persistent; bracteoles subulate, up to 1·5 mm. long, eglandular. Flowers 1·25–1·75 cm. in diameter. Sepals broadly ovate to ovate-circular, rarely oblong-ovate, 2·5–4 mm. long, ferrugineous-sericeous or glabrescent outside, with a pair of circular glands at the very base (sometimes glands much reduced) or with glands on the commissures (altogether 3–5). Petals yellow, broadly obovate to subcircular, concave (spoon-like), 6–10 mm. long, shortly clawed. Stamens with ovate-oblong anthers 4–5 mm. long; filaments rather thick, ± 1 mm. long, glabrous. Ovary densely sericeous; styles 2, 0·75–1 cm. long. Wing of samara obliquely ovate, obovate-oblong, obovate or elliptic, 1·5–2(–2·5) × 1·2–1·4(–1·7) cm., sometimes extending round nut to the base.

var. **ferrugineus** (*Engl.*) *Launert* in K.B. 19: 352 (1965). Type: Somali Republic (S.)/Kenya border area, near Uenti [Wonte], Gara Libin, *Ellenbeck* 2207 (B, holo. †); Mandera, 48 km. S. of El Wak, Bur Wein, *J. Adamson* 96 (EA, neo. !, K, isoneo. !)

Leaves ± densely ferrugineously villous or subtomentose on both surfaces, usually becoming greyish-pubescent at least beneath, rarely glabrescent. Calyx with lobes (usually only 3 or 4) biglandular at the very base. Wing of samara obliquely ovate, clasping nut nearly to the base. Fig. 1/4, p. 4.

KENYA. Northern Frontier Province: 48 km. S. of El Wak, Bur Wein, 19 Oct. 1955, *J. Adamson* 96 !
DISTR. K1; Somali Republic, Ethiopia
HAB. Semi-desert shrub, growing among rocks; ± 500 m.

SYN. *A. ferrugineus* Engl. in E.J. 36: 250 (1905); Sprague in J.B. 44: 202 (1906); Niedenzu in Arb. Bot. Inst. Akad. Braunsberg 7: 5 (1921) & in E.P. IV. 141: 263 (1928); E.P.A.: 404 (1956)

* Wrongly cited by Engler (1902) as *Riva* 206 (see K.B. 19: 351 (1965)).

5. **A. zanzibaricus** *A. Juss.*, Malpighi. Synops., in Ann. Sci. Nat., sér. 2, 13: 271 (1840) & in Archiv. Mus. Paris 3: 485 (1843); Oliv., F.T.A. 1: 279 (1868); Sprague in J.B. 44: 205 (1906); Niedenzu in Arb. Bot. Inst. Akad. Braunsberg 6: 55 (1915) & in 7: 10 (1921) & in E.P. IV. 141: 267, fig. 5/A (1928); T.T.C.L.: 296 (1949); U.O.P.Z.: 104, fig. (1949); K.T.S.: 259 (1961), pro parte. Type: Zanzibar I., *Bojer* (P, holo.)

A very attractive weak-stemmed shrub, up to 3 m. in height, rarely trailing or climbing. Very young shoots yellowish-sericeous, soon glabrescent, older ones usually densely lenticellate. Leaf-lamina oblong or obovate-oblong, (5–)6·5–11(–15) × (2–)3–4·5(–5·5) cm., often conduplicate, shortly acuminate, cuneate or rarely rounded at base, coriaceous, usually quite glabrous, with 8–12 pairs of lateral nerves, usually with 1 or 2 pairs of glands beneath near insertion of petiole; petiole robust, 0·6–1 cm. long, canaliculate, glabrous or somewhat pubescent. Inflorescences large terminal pyramidal many-flowered racemes, 8–20(–30) cm. long; axis stiff, brownish or yellowish sericeous when young, usually soon glabrescent; bracts subulate, up to 2 mm. long, persistent; bracteoles up to 1 mm. long, eglandular. Flowers 2–3 cm. in diameter. Calyx with usually 2–3 circular glands; lobes ovate, 3·5–4·5 mm. long, finely sericeous or pubescent outside, soon glabrescent. Petals bright yellow, obovate-subcircular, 1–1·25 cm. long, shortly clawed, lacerate at the margins. Stamens with linear-lanceolate or lanceolate anthers 4–6 mm. long; filaments rather thick, 1·5–2 mm. long, glabrous. Ovary ferrugineously sericeous; styles 2, up to 1·25 mm. long. Wing of samara obliquely semi-obovate (see fig. 1/8), 3–4·5 × 1·6–2·25 cm., with upper margin slightly curved or rarely straight, broadest at middle or somewhat above, obtuse or subobtuse at apex, not clasping the nutlet. Fig. 1/8, p. 4.

KENYA. Northern Frontier Province: Boni Forest, 2 Oct. 1947, *J. Adamson* 408!; Mombasa, 10 Sept. 1932, *V. G. van Someren* 1793!; Kilifi District: Mida, *Battiscombe* 801!
TANGANYIKA. Tanga District: 8 km. SE. of Ngomeni, 31 July 1953, *Drummond & Hemsley* 3567!; Uzaramo District: N. of Dar es Salaam, Kawi, July 1958, *Tweedie* 1653! & Dar es Salaam, 18 June 1931, *Musk* in *Herb. Amani* H. 5/31!
ZANZIBAR. Zanzibar I., Mangapwani, 24 Jan. 1929, *Greenway* 1146! & Chwaka, 6 Nov. 1959, *Faulkner* 2394!; Pemba I., Kengeja, 5 Mar. 1952, *R. O. Williams* 145!
DISTR. K1, 7; T3, 6; Z; P; Somali Republic
HAB. Common in coastal bushland, often on cliffs or coral outcrops, also margins of dry evergreen forest and deciduous woodland; 0–180 m.

SYN. *Anomalopteris zanzibarica* (A. Juss.) O. Ktze., Rev. Gen. Pl. 1: 87 (1891)

6. **A. chloropterus** *Oliv.*, F.T.A. 1: 279 (1868); Sprague in J.B. 44: 205 (1906); Niedenzu in Arb. Bot. Inst. Akad. Braunsberg 7: 10 (1921) & in E.P. IV. 141: 269 (1928); T.T.C.L.: 295 (1949); Launert in F.Z. 2: 111 (1963). Type: Mozambique, Zambezia, R. Chire, *Meller* (K, holo.!)

Scandent shrub or tall woody climber with branches up to 15 m. in length. Younger branches ± densely rusty pubescent, older ones quite glabrous, lenticellate. Leaf-lamina oblong or oblong-elliptic, 7–15(–19) × 2·5–5·5 (–7) cm., acute to subobtuse or acuminate, rounded, rarely subcordate or broadly cuneate, at the base, coriaceous, ferrugineous-tomentose (more densely underneath) when young, later often glabrescent except usually beside the midrib underneath, with usually 1 pair of glands beneath near insertion of petiole and with 7–13 pairs of prominent lateral nerves; petiole 2–6 mm. long, stout, canaliculate, pubescent or glabrescent. Inflorescences usually racemes terminating leafy branches, sometimes axillary, (5–)8–15 (–25) cm. long, many-flowered, pyramidal in outline; bracts triangular-lanceolate, 1–2 mm. long, densely ferrugineous-tomentose, eglandular; bracteoles very small, eglandular. Flowers ± 2·25 cm. in diameter. Calyx with 2–3 orbicular glands; lobes ovate, 3–3·5 × ± 2·5 mm., sericeous or glabrescent outside. Petals elliptic to ovate, up to 11 × 8 mm., clawed, yellow. Stamens with oblong-ovate glabrous anthers ± 4 mm. long; filaments thick, somewhat ligulate, ± 1·5 mm. long. Ovary densely ferrugineous-tomentose; styles 2, 8–10 mm. long, stout, terete, curved inwards. Wing of samara obliquely obovate, 3–4(–5) × 1·5–2·3 cm., sometimes somewhat constricted in lower half, often distally oblique-truncate or obtuse, not clasping the nut, often purplish, usually glabrous. Fig. 1/2, p. 4.

TANGANYIKA. Ulanga District: near confluence of R. Ulanga and R. Luwegu, 17 June 1932, *Schlieben* 2427!; Newala District: vicinity of R. Ruvuma, 3 Apr. 1935, *Schlieben* 6509!
DISTR. T6, 8; Malawi and Mozambique
HAB. Little known in East Africa, but recorded generally from margins and more open places in riverine and other forest, also in thickets; up to ± 250 m.

SYN. *Anomalopteris chloroptera* (Oliv.) O. Ktze., Rev. Gen. Pl. 1: 87 (1891)

7. **A. scheffleri** *Engl.* in E.J. 36: 251 (1905); Sprague in J.B. 44: 205 (1906); Niedenzu in Arb. Bot. Inst. Akad. Braunsberg 7: 12 (1921) & in E.P. IV. 141: 270 (1928); T.T.C.L.: 295 (1949). Type: Tanganyika, E. Usambara Mts., *Scheffler* 161 (B, holo. †, BM, K, iso.!)

A strong woody climber. Very young shoots ferrugineously sericeous, older ones glabrous, lenticellate. Leaf-lamina obovate, oblong-obovate or rarely elliptic, (9–)12–20(–24) × 5–7 cm., abruptly acuminate, rounded or cuneate at the base, membranous, ferrugineous-sericeous on both surfaces when young, very soon glabrescent, only the midrib beneath remaining sericeous, with up to 18 pairs of lateral nerves, usually without glands; petiole robust, 5–7 mm. long, canaliculate, ferrugineously sericeous, glabrescent. Inflorescence loose, forming axillary few- to many-flowered racemes, up to 10 cm. long; axis ± densely ferrugineous-sericeous; bracts lanceolate, 1·5–2 mm. long, acute, persistent; bracteoles very small, lanceolate, eglandular. Flowers 2·5–3 cm. in diameter. Calyx with 2 or 3 very small circular glands; lobes ovate to subcircular, 3–4 mm. long, somewhat sericeous outside, glabrescent. Petals yellow, obovate to subcircular, 8–12 × 10 mm., distinctly clawed, lacerate at margins. Stamens with lanceolate or oblong anthers 4–5 mm. long; filaments thick, ± 3 mm. long, glabrous. Ovary densely sericeous; styles ± 1 mm. long. Wing of samara obliquely semi-ovate to ovate-oblong, 5 × 2(–2·5) cm. Fig. 1/10, p. 4 (but see note below).

TANGANYIKA. E. Usambara Mts., Derema, *Scheffler* 161! & Sigi, 7 Mar. 1932, *Greenway* 2945! & Longuza, 29 May 1917, *Zimmermann* in *Herb. Amani* 6788!
DISTR. ? K7; T3; not known elsewhere
HAB. Rain-forest; 400–450 m.

NOTE. It is rather doubtful if the material cited here will continue to be retained in the same taxon. All the gatherings from the Usambara Mts. are in a very bad condition which does not allow certain determination. The plants *Greenway* 2945 and *Zimmermann* in *Herb. Amani* 6788 may represent a species distinct from *A. scheffleri*. Only more and better material can elucidate this problem. The holotype of *A. scheffleri* was destroyed in Berlin by war action; the isotypes, represented at Kew and the British Museum, are very poor; not even one flower is obtainable, nor a fruit. *Greenway* 2945 was wrongly named as *A. alopecurus*, probably because of the shape of the leaf-lamina and its size which is similar to those of that species, but it differs from *A. alopecurus* in the lack of glands associated with the bracteoles. Two further specimens may be doubtfully associated with *A. scheffleri* in a broad sense; *Drummond & Hemsley* 4331 from Kenya, Teita Hills, 8 km. NNE. of Ngerenyi (this is *A. sp.* sensu K.T.S.: 259 (1961)) and *Omari Chambo* in *Herb. Amani* 8702, from E. Usambara Mts., Kilimandege.
    A. ugandensis Sprague, at present known only from the extreme southern parts of the Sudan Republic (including areas previously within the political boundaries of East Africa), is very similar to *A. scheffleri*, differing primarily in the oblong to oblong-elliptic leaflets, usually not apiculate, more commonly rounded at the base, and the rather differently shaped samara (fig. 1/7, p. 4). A definitive treatment of these species will not be possible until much more well collected material is available for study.

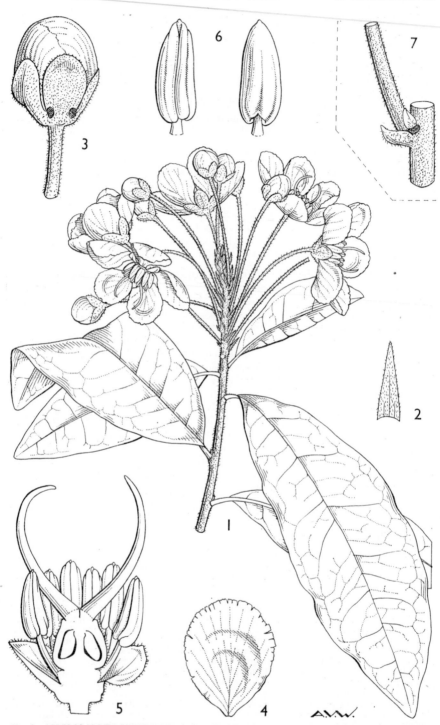

FIG. 2. *ACRIDOCARPUS CONGESTUS*—**1**, flowering branch, × 1; **2**, bract, × 6; **3**, bud, × 4; **4**, petal, × 2; **5**, longitudinal section of flower, × 4; **6**, anthers, × 6. *A. SMEATHMANII*—**7**, part of inflorescence-axis, showing gland at base of bracteole, × 6. 1–6, from *Wallace* 453; 7, from *Procter* 707.

8. **A. pauciglandulosus** *Launert* in K.B. 19 : 350 (1965). Type : Tanganyika, Lindi District, Rondo [Mwera] Plateau, *Schlieben* 5498* (BM, holo.!, M, iso.!)

A straggling shrub (or ? woody climber) with branches up to 3 m. in length. Younger branches ± densely ferrugineous-pubescent, older ones glabrous, lenticellate. Leaf-lamina oblanceolate to (rarely) obovate, 4–7 × 1·5–2·5(–3) cm., obtuse or shortly acuminate, cuneate, coriaceous, with margins slightly involute, finely ferrugineous-pubescent when young, soon glabrescent or sometimes upper surface and midrib beneath remaining finely pubescent, with 9–13 pairs of lateral nerves, eglandular or with a pair of very small glands at the very base beneath; petiole 1–2 mm. long, ± robust, sub-canaliculate, pubescent or glabrous. Inflorescences mostly axillary, 3–5 cm. long, many-flowered, with a fairly stout rhachis, ± densely ferrugineous-sericeous, forming pyramidal or corymbiform racemes; bracts triangular-subulate, 1·5–2·5 mm. long, persistent; bracteoles subulate, 1–1·5 mm. long, eglandular. Flowers ± 2·25 cm. in diameter. Calyx eglandular or with 1–2(–3) small circular glands, ± densely sericeous outside; lobes 2–2·5 (–3) mm. long, ovate to ovate-oblong, usually glabrous but shortly ciliate at margins. Petals yellow, obovate, ± 12 mm. long, entire or shortly lacerate, clawed. Stamens with oblong-ovate anthers 3–4 mm. long; filaments cylindric to ligulate, 1–1·5 mm. long, carnose, glabrous. Ovary densely sericeous. Styles 2, 7–10 mm. long, usually glabrous. Wing of samara obliquely obovate, 4–5 × 1·75–2·25 cm., narrowed in lower half, broadest across the middle, distally oblique-truncate and obtuse (see fig. 1/1), clasping the nut only by a very small ridge, usually glabrous, sometimes finely pubescent. Fig. 1/1, p. 4.

TANGANYIKA. Lindi District: Rondo [Mwera] Plateau, 21 Nov. 1934, *Schlieben* 5498! Newala District: Kitangari, 23 Mar. 1943, *Gillman* 1329!
DISTR. **T**6, 8; not known elsewhere
HAB. Little known, but recorded from thickets; ± 200–600 m.

SYN. *A. natalitius* A. Juss. var. *acuminatus* Niedenzu in Arb. Bot. Inst. Akad. Brauns-berg 7: 8 (1921) & in E.P. IV. 141: 267 (1928), pro parte excl. specim. Afr.–austr.; T.T.C.L.: 295 (1949). Types: Tanganyika, Uzaramo District, Pugu Hills, *Holtz* 467 (B, syn. †) & Uzaramo District, without precise locality, *Goetze* 13 (B, syn. †)

9. **A. congestus** *Launert* in K.B. 19: 349 (1965). Type: Tanganyika, Uluguru Mts. above Morningside, *Eggeling* 6768 (K, holo.!, EA, FHO, PRE, iso.!)

A lofty woody climber. Younger branches ± densely ferrugineous-seri-ceous, older ones usually glabrous, lenticellate. Leaf-lamina obovate to oblong-obovate, 4·5–9(–11) × 2–4 cm., obtuse or acuminate, cuneate, with margins somewhat involute, subcoriaceous, glabrous, with 12–16 pairs of lateral nerves, usually eglandular (rarely with a pair of much reduced glands beneath near insertion of petiole); petiole 3–5 mm. long, usually glabrous, canaliculate. Inflorescences on leafy branches terminating in racemes, many-flowered, with a robust densely ferrugineous-sericeous rhachis; bracts lanceolate-triangular, 3–4 mm. long, acute, persistent; bracteoles triangular, 1–2 mm. long, eglandular. Flowers 2·5–3 cm. in diameter. Calyx with (1–)2(–3) circular glands, rarely eglandular; lobes ovate to oblong-ovate, 5–6 mm. long, ferrugineous-sericeous or glabrescent outside, shortly ciliate at margins. Petals canary yellow, subcircular to broadly obovate, 11–15 mm. long, deeply lacerate at margins or rarely entire, very

* There is another specimen bearing the same number but with a different locality on its label ("Lutamba See, 40 km. westlich Lindi, 1934–35") in Berlin, thus it is doubtful if it represents the same gathering!

shortly clawed. Stamens with oblong-ovate anthers 4·5–5 mm. long; filaments 2–3 mm. long, flattened, enlarged at base, glabrous. Ovary densely
ferrugineous-sericeous; styles 2, ± 12 mm. long, usually glabrous. Wing of
samara obliquely oblong to oblong-obovate, 4·5–6 × 1·8–2·25 cm., with upper
edge curved towards the acute or subacute apex, clasping the nut to a certain
extent but never to its base (see fig. 1/6), usually glabrous. Figs. 1/6, p. 4,
2/1–6, p. 10.

TANGANYIKA. Morogoro District: without precise locality, 18 Nov. 1932, *Wallace*
    453 ! & Uluguru Mts., Bondwa Mts. above Morningside, Dec. 1953, *Eggeling* 6768 !
DISTR. T6; not known elsewhere
HAB. Rain-forest; 1350–1620 m.

SYN. *A. sp.* sensu T.T.C.L.: 296 (1949)

2. TRISTELLATEIA

Thou., Gen. Nov. Madag. 14, No. 47 (1806)

Woody climbers, usually glabrous. Leaves opposite, entire, usually with
2 glands on margin of lamina near base (in *T. africana* glands near apex of
petiole); stipules very small. Inflorescence racemose, terminal; pedicels
articulated, 2-bracteolate. Flowers actinomorphic, bisexual. Sepals persistent, in some species with dorsal glands. Petals oblong, clawed, keeled,
entire. Stamens with glabrous basifixed anthers; filaments of outer whorl
longer and broader at base. Ovary globose; only 1 (very seldom 2) styles
fully developed, terete. Fruit subligneous; lateral wing divided into 4–10
narrow stellately arranged lobes; median wing usually absent; samara
sometimes with a dorsal crest. Seeds subglobose, with a short acumen formed
by the radicle.

A genus of 22 species, mainly in Madagascar, but with one species known from the
African continent and one from SE. Asia.

NOTE.   *T. australasiae* A. Rich. in Sert. Astrolabi., t. 15 (1833) is a species cultivated in
    the Victoria Gardens of the Zanzibar Residence (see U.O.P.Z.: 477 (1949)). From
    T. africana it can be easily distinguished by having the glands on the margins of the
    leaf-laminae near the insertion of the petiole.

T. africana *S. Moore* in J.B. 15: 289 (1877); Niedenzu in E.P. IV. 141: 64
(1928); Arènes in Mém. Mus. Nat. Hist. Paris, n. sér., 21: 311 (1947);
T.T.C.L.: 297 (1949); E.P.A.: 404 (1956); K.T.S.: 260 (1961); Launert
in Bol. Soc. Brot., sér. 2, 35: 49, t. 7 (1961) & in F.Z. 2: 113, t. 14 (1963).
Type: Kenya, coast near Mombasa, *Hildebrandt* 1974 (K, holo.!, BM, iso.!)

Woody climber up to 5 m. or more in length; younger stems usually
greyish-pubescent, older ones glabrous, densely lenticellate. Leaf-lamina
ovate or broadly elliptic, 4·5–9·5 × 3·5–5 cm., acute or obtuse, rounded or
cordate at the base, coriaceous to chartaceous, glabrous (younger ones
sometimes ± sericeous); petiole 1·5–3 cm. long, with a pair of glands near
apex. Inflorescence 5–12 cm. long, many-flowered; pedicels 5–18 mm. long,
decussate, distinctly articulate; bracts linear-lanceolate, 2–3 mm. long;
bracteoles very short, subulate. Flowers 2·3–2·5 cm. in diameter. Sepals
oblong, 4–5 mm. long, sericeous outside. Petals bright yellow, oblanceolate
to oblong, 10–12 mm. long, rounded at apex, cordate or subsagittate at base,
shortly clawed. Anthers linear, 3–4 mm. long, orange; filaments glabrous,
somewhat incurved. Samara 1·5–2 cm. in diameter; lateral wing divided
into 6 oblanceolate-linear usually denticulate lobes; dorsal crest with a spine
8–10 mm. long. Fig. 3.

KENYA.   Kwale District: 29 km. S. of Mombasa, Jardini beach, 26 Aug. 1953, *Drum-
    mond & Hemsley* 3980!; Mombasa, Feb. 1932, *V. G. van Someren* 1791!; Kilifi
    District: R. Sabaki N. of Malindi, 2 Nov. 1961, *Polhill & Paulo* 688!

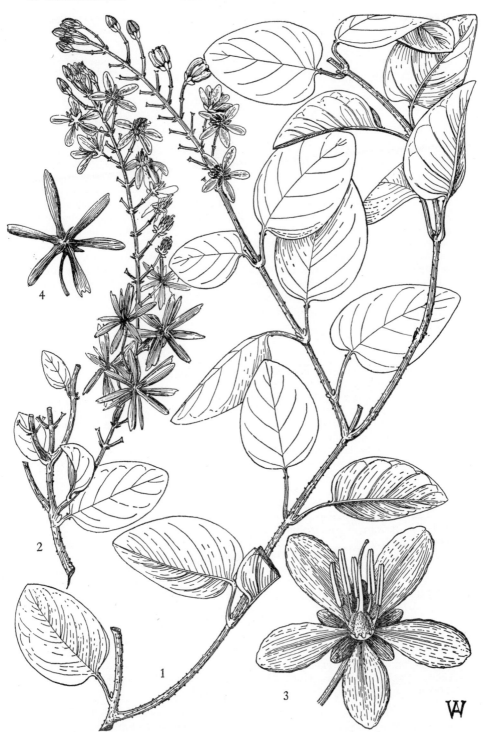

FIG. 3. *TRISTELLATEIA AFRICANA*—**1,** flowering branch, × ⅔; **2,** same with fruits also, × ⅔; **3,** flower, × 3; **4,** fruit, × 1. All from *Exell, Mendonça & Wild* 685. Reproduced by permission of the Editors of " Flora Zambesiaca ".

TANGANYIKA.   Tanga District: N. of Tanga, 7 Jan. 1939, *Greenway* 5826!; Pangani
   District: Msubugwe Forest Reserve, Sept. 1955, *Semsei* 2273!; Lindi, Mar. 1952,
   Semsei 708!
DISTR.   **K**7; **T**3, 6, 8; Mozambique*
HAB.   Coastal bushland, thicket and mangrove swamp, often on coral outcrops or
   near high-tide mark, also in dry evergreen forest and wooded grassland; 0–50 m.

3. TRIASPIS

Burch., Trav. Int. S. Afr. 2: 280, t. 290 (1824)

Small trees, scandent or semi-scandent shrubs to woody climbers. Leaves
opposite, subopposite or rarely ternate, usually with 2–4 glands on undersurface near base, petiolate or sessile, with or without interpetiolar stipules.
Inflorescences terminal or axillary, usually forming many-flowered corymbs,
umbels or panicles; bracts and bracteoles usually present, usually deciduous.
Flowers actinomorphic or zygomorphic. Sepals 5, almost always without
glands. Petals 5, clawed, usually with fringed or denticulate margins;
pedicels as long as or longer than the peduncle and articulated with it.
Stamens 10; anthers basifixed, usually glabrous; filaments subulate, glabrous
or farinose-pubescent. Ovary hairy or glabrous, 3-locular (or sometimes
2-locular outside East Africa); styles (2–)3, somewhat curved, with incurving
stigmas. Samara with a circular or ovate membranous or coriaceous lateral
wing; dorsal wing shorter and narrower or absent.

An African genus of 15 species.

Styles 2; flowers actinomorphic:
   Leaves from a cuneate base elliptic, elliptic-oblong,
      lanceolate-oblong, rarely ovate-oblong, (2·5–)3–
      4·5(–6) × (1·3–)1·8–2·5(–3·5) cm., obtuse, usually
      glabrous on both surfaces; secondary nerves
      usually ± prominent beneath   .   .   .   1. *T. erlangeri*
   Leaves from a usually rounded rarely cuneate base
      ovate or broadly elliptic, rarely ovate-lanceolate,
      1·5–2·75 (rarely –7) × 0·9–1·8 (rarely –3) cm.,
      obtuse or subacute, ± densely pubescent or
      canescent (more densely beneath), older ones
      rarely glabrescent; secondary nerves usually
      indistinctly visible   .   .   .   .   2. *T. niedenzuiana*
Styles 3; flowers zygomorphic:
   Ovary glabrous .   .   .   .   .   .   3. *T. schliebenii*
   Ovary densely hairy (usually tomentose):
      Filaments glabrous; stipules very small or usually
         absent   .   .   .   .   .   .   4. *T. mozambica*
      Filaments finely pubescent; stipules fairly large
         (3–4·5 mm. long)   .   .   .   .   5. *T. macropteron*

1. **T. erlangeri** *Engl.* in E.J. 36: 248 (1905); Chiov., Result. Sci. Miss.
Stef.-Paoli Som. Ital. 1: 36 (1916); Niedenzu in Verz. Vorl. Akad. Braunsberg S.-Sem. 1924: 5 (1924) & in E.P. IV. 141: 42 (1928); E.P.A.: 403
(1956). Types: Ethiopia, Harar Province, *Ellenbeck* 998 & 1034 (B, syn. †);
Gamu Gofa, Gondaraba, *Corradi* 6975 (FI, neo.!)

Small much branched erect shrub, up to 2 m. or more high, with branchlet-
tips twining; younger stems canescent from appressed or somewhat spread-

* Possibly also in Somali Republic (S.), from where Chiovenda described *Tristellateia
somalensis*, later reduced to *T. africana* var. *somalensis* (Chiov.) Arènes. The type,
Paoli 101 from Mogadishu, has not however been examined.

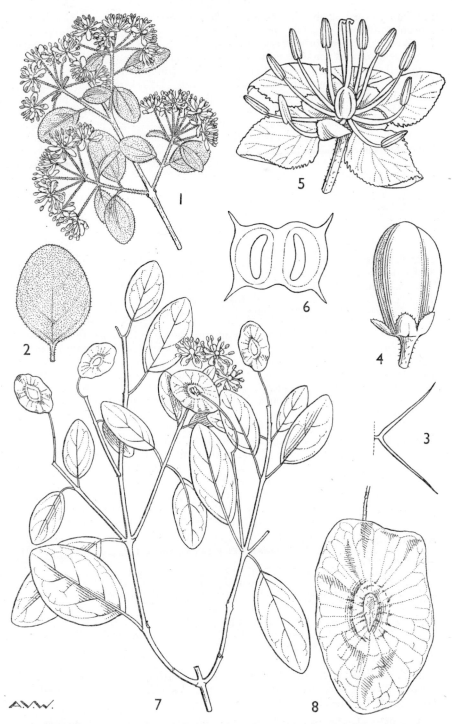

FIG. 4. *TRIASPIS NIEDENZUIANA*—**1,** flowering branch, × 1; **2,** leaf, × 1; **3,** hair, × 40; **4,** bud, × 8; **5,** flower, × 6; **6,** ovary in transverse section, × 20. *T. ERLANGERI*—**7,** fertile branch, × 1; **8,** fruit, × 2. 1–3, 5, 6, from *Gillett* 12711; 4, from *Dummer* 5018; 7, from *Napier* 1036; 8, from *Corradi* 6975.

ing hairs, very soon glabrescent. Leaf-lamina elliptic, elliptic-oblong, lanceolate-oblong, rarely ovate-oblong, $(2\cdot5-)3-4\cdot5(-6) \times (1\cdot3-)1\cdot8-2\cdot5$ $(-3\cdot5)$ cm., obtuse or subobtuse, cuneate, subcoriaceous, usually with involute edges, concolorous, pale green to yellowish when dried up; petiole 3–9 mm. long, pubescent or glabrous, slightly canaliculate. Flowers in 4–8-flowered loose corymbs terminating annual branches, $1\cdot25-1\cdot5$ cm. in diameter; bracts linear, 3–5 mm. long. Sepals ovate or oblong-ovate, ± 2 mm. long. Petals white, pinkish, or white and tinged with pink, broadly elliptic or elliptic-oblong, 4–5 mm. long, shortly clawed. Anthers broadly elliptic, $0\cdot8-1$ mm. long; filaments thread-like, $3\cdot5-4\cdot2$ mm. long. Styles ± 5 mm. long, glabrous. Samara elliptic or elliptic-oblong, $1\cdot5-3$ cm. long and $1\cdot1-2$ cm. broad, entire or slightly retuse at apex; dorsal wing reduced to a tiny crest or absent. Fig. 4/7, 8, p. 15.

KENYA. Teita District: Voi, 11 May 1931, *Napier* 1036! & Voi, Mzinga Hill, 11 Jan. 1964, *Verdcourt* 3893!
DISTR. **K7**; Ethiopia, Somali Republic
HAB. Rocky hills with *Euphorbia, Commiphora* bushland; 600 m.

2. **T. niedenzuiana** *Engl.* in E.J. 36: 247 (1905); Chiov., Result. Sci. Miss. Stef.-Paoli Som. Ital. 1: 36 (1916); Niedenzu in Verz. Vorl. Akad. Braunsberg S.-Sem. 1924: 5 (1924) & in E.P. IV. 141: 44, fig. 11/A–C (1928); T.T.C.L.: 297 (1949); E.P.A.: 404 (1956). Type: Tanganyika, foot of NW. Pare Mts., *Uhlig* 863 (B, syn. †, EA, isosyn. !)

A small attractive semi-erect or scandent shrub (sometimes twining with ends of branches), up to 3 m. or more high; younger stems and inflorescences $\pm$ densely covered with somewhat stiff short hairs, older ones glabrescent. Leaves petiolate; lamina from a usually rounded rarely cuneate base ovate or broadly elliptic, rarely ovate-lanceolate, $1\cdot5-2\cdot5(-7) \times 0\cdot9-1\cdot8(-3)$ cm., obtuse or subacute, older ones $\pm$ densely pubescent (more so beneath); secondary nerves usually indistinctly visible; petiole 2–4(–6) mm. long, densely pubescent. Flowers in loose few-flowered corymbs terminating leafy annual branches, $1-1\cdot25$ cm. in diameter; bracts linear, 2–4 mm. long. Sepals ovate, ± 2 mm. long. Petals pure white, lilac or pinkish, oblong or oblong-ovate, $\pm$ spoon-shaped, $3\cdot5-4$ mm. long, shortly fimbriate at edges near base, clawed. Stamens with anthers $1-1\cdot3$ mm. long; filaments 3–5 mm. long. Styles $\pm 4\cdot5-5\cdot5$ mm. long. Samara elliptic or ovate-elliptic, $1\cdot8-2 \times 0\cdot9-1\cdot1$ cm. (in available material, but no fruits were fully developed); lateral wing distinctly retuse at apex; dorsal wing reduced. Fig. 4/1–6, p. 15.

KENYA. Northern Frontier Province: Moyale, 2 Nov. 1952, *Gillett* 14114!; Machakos District: Mtito Andei, *Battiscombe* 920!; Kwale District: Taru plains, 28 Mar. 1901, *Kassner* 533!
TANGANYIKA. Pare Mts. at NW. foot, 12 Dec. 1901, *Uhlig* 863!
DISTR. **K1, 4, 7**; **T3**; Ethiopia, Somali Republic
HAB. Deciduous bushland, sometimes on stony hillsides or rock outcrops, also degraded " montane scrub "; 500–1350 m.

SYN. *Tristellateia cynanchoïdes* Chiov., Fl. Somala 2: 44 (1932); E.P.A.: 404 (1956). Type: Somali Republic (S.), Uegit [" Meggit "], *Senni* 837 & Oddur, *Senni* 805 bis (both FI, syn.)

3. **T. schliebenii** *A. Ernst* in N.B.G.B. 12: 709 (1935); T.T.C.L.: 297 (1949). Type: Tanganyika, Lindi District, Lake Lutamba, *Schlieben* 6093 (B, holo. †, BM, iso. !)

A woody slender-stemmed climber; younger stems and inflorescences appressed yellowish-sericeous, older ones glabrescent and lenticellate. Leaf-lamina ovate-oblong or lanceolate, $5\cdot5-12 \times 1\cdot5-4$ cm., acuminate, base rounded or sometimes slightly cordate, chartaceous, hirsute on both surfaces, more densely so beneath; petiole 4–12 mm. long, canaliculate, hirsute; stipules very small, deciduous. Flowers in terminal and axillary many-flowered corymbs, zygomorphic, ± 1 cm. in diameter; bracts lanceolate, up to 2 mm. long; bracteoles filiform, ± 1 mm. long. Sepals ovate-oblong, 1–1·3 mm. long, glabrous. Petals (? white), oblong-ovate, ± 9 mm. long and 4 mm. broad, subhastate, entire or irregularly dentate at margins, shortly clawed. Stamens with anthers ± 1·5 mm. long; filaments filiform, ± 2 mm. long. Ovary glabrous; styles ± 4 mm. long, glabrous. Samara unknown.

TANGANYIKA. Lindi District: Lake Lutamba, 7 Mar. 1935, *Schlieben* 6093 !
DISTR. **T**8 (known only from the type-gathering)
HAB. " Climber on shrubs and small trees in forest "; 200 m.

4. **T. mozambica** *A. Juss.*, Malpighi. Synops., in Ann. Sci. Nat., sér. 2, 13 : 268 (1840) & in Archiv. Mus. Paris 3 : 505 (1843); Oliv., F.T.A. 1 : 281 (1868); Engl., P.O.A. A : 76 (1895), in obs.; Niedenzu in Arb. Bot. Inst. Akad. Braunsberg 6 : 23 (1915) & in Verz. Vorl. Akad. Braunsberg S.-Sem. 1924 : 6 (1924) & in E.P. IV. 141 : 46 (1928); T.T.C.L. : 297 (1949); Launert in F.Z. 2 : 117 (1963). Type: Mozambique, Delagoa Bay, *Forbes* (K, holo. !)

Small climber, up to 3 m. long, with younger stems and inflorescences ± densely ferrugineous-sericeous, older ones glabrescent. Leaf-lamina lanceolate, ovate, ovate-lanceolate or elliptic, 2·5–8 × 2·4 cm., obtuse or acute, apiculate, base rounded or subcordate, membranous, ferrugineous-sericeous on both surfaces when young, soon glabrescent; petiole 0·7–1·2 cm. long, canaliculate, usually sericeous; stipules very small, deciduous. Flowers numerous in ± loose, terminal or axillary corymbs, ± 1–1·5 cm. in diameter, zygomorphic; bracts linear-lanceolate, 3·5 mm. long; bracteoles linear, 1–2 mm. long. Sepals ovate, ± 2 mm. long, sericeous outside. Petals green or yellowish-green, obovate and ± cucullate, 4·5–6 mm. long, spreading or recurved, shortly clawed, with margins shortly fimbriate or crenulate. Stamens with ovate or oblong-ovate anthers 1–1·25 mm. long; filaments 2·5–3 mm. long, glabrous. Ovary densely sericeous; styles glabrous. Samara with lateral wing subcircular, 2–2·7 cm. in diameter, retuse or ± emarginate at apex; dorsal wing crest-like, 0·8 × 0·25 cm., subcordate at base.

KENYA. Northern Frontier Province: Boni Forest, Mararani, 6 Sept. 1961, *Gillespie* 309 !; Kwale District: Shimoni, 20 Aug. 1953, *Drummond & Hemsley* 3910 !; Kilifi District: Kibarani, 29 Jan. 1946, *Jeffery* 451 !
TANGANYIKA. Moshi District: 14·5 km. E. of Moshi, 4 Nov. 1955, *Milne-Redhead & Taylor* 7225 !; E. Usambara Mts., Kisiwani, 13 Jan. 1937, *Greenway* 4820 !; Pangani District: Bushiri Estate, 4 Sept. 1950, *Faulkner* 679 !
ZANZIBAR. Zanzibar I., Haitajwa, 2 July 1930, *Vaughan* 1391 ! & 25 May 1935, *Vaughan* 2233 !
DISTR. **K**1, 7; **T**2, 3, 6; **Z**; Mozambique
HAB. Coastal and riverine bushland or thicket, also dry evergreen forest and lowland rain-forest; 0–800 m.

SYN. *T. mozambica* A. Juss. forma *lanceolata* Niedenzu in Arb. Bot. Inst. Akad. Braunsberg 6 : 23 (1915) & in E.P. IV. 141 : 47 (1928); T.T.C.L. : 297 (1949). Type: Tanganyika, E. Usambara Mts., Mashewa, *Holst* 3565 (B, holo. †, BM, K, iso. !)
    T. mozambica A. Juss. forma *subcordata* Niedenzu in Arb. Bot. Inst. Akad. Braunsberg 6 : 23 (1915) & in Verz. Vorl. Akad. Braunsberg S.-Sem. 1924 : 6 (1924) & in E.P. IV. 141 : 47 (1928); T.T.C.L. : 297 (1949). Type: Tanganyika, Bagamoyo, *Hildebrandt* 1297 (B, holo. †)
    T. mozambica A. Juss. forma *gracilis* Niedenzu in Verz. Vorl. Akad. Braunsberg S.-Sem. 1924 : 6 (1924) & in E.P. IV. 141 : 47 (1928); T.T.C.L. 2 : 297 (1949). Type: Tanganyika, Pangani, *Stuhlmann*, ser. I, 548 (B, holo. †)

VARIATION. The leaves of this species show a remarkable range of variation in shape and size, but it is impossible to maintain the forms described by Niedenzu based on those characters, because the variation can often be found in the same specimen.

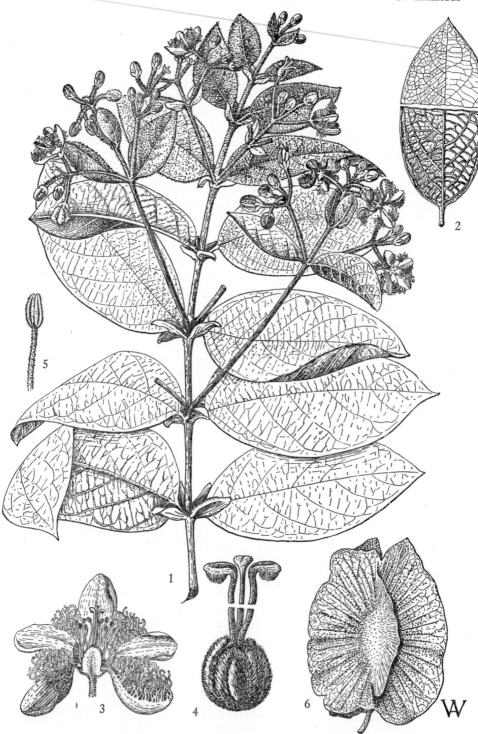

FIG. 5. *TRIASPIS MACROPTERON* subsp. *MASSAIENSIS*—**1,** flowering branch, × ⅔; **2,** leaf, showing both surfaces, × ⅔; **3,** flower, × 2; **4,** ovary, × 8; **5,** stamen, × 8; **6,** fruit, × ⅔. All from *Chase* 5200. Reproduced by permission of the Editors of " Flora Zambesiaca ".

5. **T. macropteron** *Oliv.*, F.T.A. 1: 281 (1868); Launert in F.Z. 2: 115 (1963). Type: Angola, Cuanza Norte, *Welwitsch* 1039 (LISU, syn., BM, K, isosyn. !)

A tall woody climber (sometimes creeping) up to 4 m. or more in length; stems densely ferrugineous-pubescent when young, becoming glabrous when older. Leaf-lamina ovate, broadly lanceolate, lanceolate-oblong or oblong, 6–13 × 2–6·5 cm., acute or apiculate, or sometimes subobtuse, cordate, rounded or cuneate at base, densely ferrugineous-pubescent on both surfaces when young, later usually glabrous, dark green above, grey-green beneath, with the secondary and tertiary nerves ± strongly developed; petiole 0·5–2 cm. long, canaliculate, usually glabrous; stipules broadly lanceolate, obovate or oblong, ± elliptic, 2–15 × 1–8 mm., deciduous or ± persistent, usually glabrous. Flowers many in terminal or axillary corymbose panicles, 1·5–2·5 cm. in diameter, zygomorphic; bracts 0·5–2·5 mm. long, deciduous; bracteoles ± 1 mm. long, ± persistent. Sepals oblong-elliptic, 2–3 × 1–1·5 mm., sericeous outside, glabrescent. Petals white to cream or yellow-orange (? red), 6–9 × 4–5 mm., distinctly clawed, deeply fimbriate at margins. Stamens with ovate-oblong anthers 1–1·5 mm. long; filaments 4–7 mm. long, minutely pubescent. Ovary densely sericeous; styles 4–8 mm. long, glabrous. Samara with lateral wing circular or broadly ovate, 4–5 cm. in diameter, glabrous, somewhat undulate; dorsal wing crest-like, 1·5–3 × 0·4–0·6 cm.

subsp. **massaiensis** (*Niedenzu*) *Launert* in Bol. Soc. Brot., sér. 2, 35: 31 (1961) & in F.Z. 2: 117, t. 15 (1963). Type: Tanganyika, Mwanza District, Kayenzi [Kagehi], *Fischer* 66 (K, isosyn. !)

Leaf-lamina broadly lanceolate, lanceolate-oblong or oblong, base cuneate, rarely rounded or very slightly cordate; tertiary nerves strongly developed; stipules broadly lanceolate to obovate, 3–15 × 3–8 mm., ± persistent. Fig. 5.

TANGANYIKA. Shinyanga, 20 May 1931, *B. D. Burtt* 2416!; Ufipa District: escarpment above Kasanga, 30 Mar. 1959, *Richards* 11002!; W. Mpwapwa, 27 Dec. 1931, *Hornby* 434!
DISTR. T1, 2, 4, 5, 7; Mozambique, Malawi, Zambia and Rhodesia
HAB. Deciduous woodland, bushland and thicket, often riparian or on rocky hills; 900–1650 m.

SYN. *T. speciosa* Niedenzu in P.O.A. C: 232 (1895); Engl. in E.J. 28: 416 (1900). Types: Tanganyika, Mwanza District, Kayenzi [Kagehi], *Fischer* 77 & 290 & Mwanza, *Stuhlmann* 4575 & Karumo, *Stuhlmann* 3577 (all B, syn. †, K, isosyn. of *Fischer* 290 !)
T. stipulata Engl. in E.J. 43: 382 (1909), *non* Oliv. (1868), *nom. illegit.* Type: Tanganyika, Dodoma District, Kilimatinde, *Claus* (B, holo. †)
T. macropteron Oliv. var. *speciosa* (Niedenzu) Niedenzu forma *brevistipulata* Niedenzu in Verz. Vorl. Akad. Braunsberg S.-Sem. 1924: 7 (1924) & in E.P. IV. 141: 52 (1928), pro parte, *nom. illegit.* Types: as for *T. speciosa*
T. macropteron Oliv. var. *speciosa* (Niedenzu) Niedenzu forma *massaiensis* Niedenzu in Verz. Vorl. Akad. Braunsberg S.-Sem. 1924: 7 (1924) & in E.P. IV. 141: 52 (1928)

NOTE. Subsp. *macropteron* of NE. Angola and western Congo Republic, differs from subsp. *massaiensis* mainly in having leaves with a usually cordate base and less strongly developed tertiary nerves underneath, also very small caducous stipules.

4. FLABELLARIOPSIS

Wilczek in B.J.B.B. 25: 303 (1955) & emend. in B.J.B.B. 29: 193 (1959)

Woody climbing plants with lenticellate stems. Leaves opposite; stipules present, linear-subulate, deciduous. Flowers in few-flowered lax umbellate or panicled racemes, regular, white, pedicellate. Bracts and bracteoles persistent. Sepals 5, imbricate, shortly connate at the base, without glands, closed over the petals in bud. Petals 5, free, clawed, deciduous. Stamens 10,

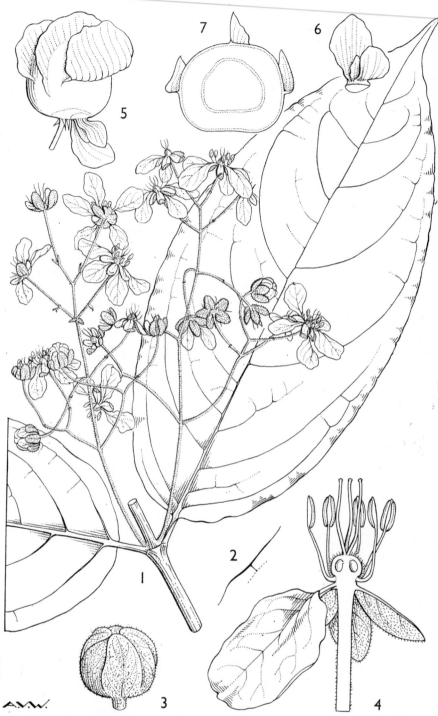

Fig. 6. *FLABELLARIOPSIS ACUMINATA*—**1,** flowering branch, × 1; **2,** hair, × 20; **3,** bud, × 4; **4,** flower in transverse section showing one petal in surface view, × 4; **5,** fruit, × 1; **6,** suppressed mericarps of the fruit, × 1; **7,** fruit in transverse section, × 1. 1–4, from *Ghesquière* 4254; 5–7, from *Pierlot* 1382 (after J. Lerinckx, fig. 31 in B.J.B.B. 29: 193 (1959)).

all fertile; anthers basifixed, with a cordate base; filaments short, free or somewhat connate at the very base, glabrous. Ovary 3-locular, tomentose; styles 3, filiform, elongate, with terminal very small stigmas. Samara subglobose, drupe-like, with only 1 of the mericarps fully developed, sublignose, with nearly parallel crest-like coriaceous free wings, 1 on the apex, the others laterally.

A monospecific genus confined to tropical Africa.

F. acuminata (*Engl.*) *Wilczek* in B.J.B.B. 25: 304, t. 8 (1955) & in F.C.B. 7: 217, t. 24 (1958) & in B.J.B.B. 29: 193, fig. 31 (1959). Types: Tanganyika, Iringa District, Mt. Uzungwa [Utschungwe], *Goetze* 610 (B, holo. †); E. Usambara Mts., near Amani, Bomole, *Zimmermann* 856 (BM, neo.!, PRE, isoneo.)

A strong woody climber up to 6 m. or more in length; stems terete, lenticellate, usually glabrous. Leaf-lamina varying from elliptic-lanceolate to oblong or oblong-elliptic, 6–13 × 2·5–6·5 cm., ± abruptly apiculate or acuminate, broadly cuneate, rounded or subcordate at the base, subcoriaceous, usually glabrous on both surfaces (rarely pubescent when young), with 5–9 pairs of lateral nerves and with 2–4 circular glands near the margin at the base beneath; petiole 0·5–1·5 cm. long, canaliculate, sparsely pubescent or glabrous; stipules ± 1 mm. long, linear-subulate, soon deciduous. Inflorescences pubescent to sometimes tomentose, few-flowered; bracts oblanceolate or elliptic, 3·5–5 mm. long, pubescent, usually persistent; bracteoles very small, deciduous. Flowers 1–1·5 cm. in diameter, whitish or yellowish; pedicels up to 20 mm. long, slender. Sepals oblong-elliptic, (2–)3–6 mm. long, tomentose outside, pubescent inside. Petals obovate or broadly elliptic, 5–8 mm. long, entire but sometimes crenulate or ciliolate at the margins; claw 3–4·5 mm. long. Anthers oblong, 1·5–2 mm. long, usually ± pilose; filaments 2–3 mm. long, glabrous. Ovary tomentose; styles filiform, 2·5–4 mm. long. Samara 2·4 2·7 cm. in diameter; apical wing 8–10 mm. broad; lateral wings 2·8–3 × 1–1·4 cm.; often all wings lacking especially in very mature fruits. Fig. 6.

UGANDA. Ankole District: Kalinzu Forest, Aug. 1936, *Eggeling* 3202!
TANGANYIKA. W. Usambara Mts., Matondwe Hill above Kwai, 29 May 1953, *Drummond & Hemsley* 2809!; Ulanga District: vicinity of Mahenge, 10 Dec. 1931, *Schlieben* 1542!; Iringa District: Mufindi, 1 Oct. 1934, *R. M. Davies* 919!
DISTR. **U2**; **T3**, 6, 7; Congo Republic
HAB. Rain-forest, dry evergreen and riverine forest, also wooded grassland; 50–2000 m.

SYN. *Triaspis acuminata* Engl. in E.J. 28: 416 (1900)
    Brachylophon niedenzuianum Engl. in Arb. Bot. Inst. Akad. Braunsberg 6: 47 (1915) & in E.P. IV. 141: 249 (1928). Type: Tanganyika, W. Usambara Mts., Kwai, *Albers* 322 (B, holo. †)
    B. acuminatum (Engl.) Niedenzu in E.P. IV. 141: 250 (1928), in syn.; T.T.C.L.: 296 (1949)
    B. niedenzuianum Engl. var. *acuminatum* (Engl.) Niedenzu in E.P. IV. 141: 250 (1928), *comb. illegit.*
    [*Triaspis lateriflora* sensu Staner in Ann. Soc. Sci. Brux., sér. 2, 40: 37 (1938); T.T.C.L.: 296 (1949), *non* Oliv.]
    Brachylophon acuminatum (Engl.) Niedenzu var. *niedenzuianum* (Engl.) Brenan, T.T.C.L.: 296 (1949)

5. FLABELLARIA

Cavan., Diss. 9: 436, t. 264 (1790); Hook. f. in G.P. 1: 259 (1862); Niedenzu in E.P. IV. 141: 38 (1928)

Woody climbers. Leaves opposite, petiolate, without stipules. Flowers regular, in many-flowered terminal or axillary panicled racemes, pedicellate,

FIG. 7. *FLABELLARIA PANICULATA*—**1**, flowering branch, × 1; **2**, bud, × 6; **3**, flower, × 6; **4**, fruit,
× 1. 1–3, from *Gillman* 463; 4, from *Sillitoe* 339.

white or cream. Sepals 5, valvate, closed over petals in bud, eglandular. Petals 5, not clawed, entire, glabrous, oblong-lanceolate to sometimes oblanceolate. Stamens 10, all bearing anthers; filaments free or somewhat connate at the very base; anthers elliptic or oblong, basifixed. Ovary 3-locular, densely pilose; styles 3, much longer than the stamens. Samara with 2 lateral wings which are connate at the base and distinct at the top.

A monospecific genus confined to tropical Africa.

F. paniculata Cavan., Diss. 9: 436, t. 264 (1790); Oliv., F.T.A. 1: 282 (1868); Engl. in Z.A.E. 2: 435 (1912) & in V.E. 3(1): 826, fig. 391 (1915); Niedenzu in E.P. IV. 141: 38, t. 10 (1928); T.T.C.L.: 296 (1949); Exell & Mendonça in C.F.A. 1: 250 (1951); F.P.S. 2: 44 (1952); F.W.T.A., ed. 2, 1: 353 (1958); Wilczek in F.C.B. 7: 215, fig. 3 (1958). Type: Sierra Leone, *Smeathman in Herb. Thouin**

A tall climber up to 15 m. in length; stems up to 10 cm. or slightly more in diameter, lenticellate, younger ones with a ± dense grey or yellowish silky indumentum. Leaf-lamina broadly elliptic, ovate, ovate-subcircular, or rarely lanceolate, 5–15 × 4–10 cm., obtuse, subacute or apiculate, rounded or subcordate at the base, subcoriaceous, upper surface usually glabrous, lower surface appressed silky-tomentose; lateral nerves 4–6 pairs; petiole 1–2·5 cm. long, canaliculate, tomentose. Inflorescences up to 20 cm. long, ± lax; bracts oblanceolate, 3–7 mm. long, deciduous or ± persistent; bracteoles very small, elliptic, ± persistent; pedicels up to 5 mm. long. Flowers ± 1 cm. in diameter. Sepals oblong-lanceolate, 5–6·5 mm. long, usually reflexed, tomentose outside. Petals oblanceolate, up to 7 mm. long, entire, glabrous, rounded at the apex. Anthers elliptic to oblong-elliptic, 1·25–1·75 mm. long; filaments 2–3 mm. long, glabrous. Styles 3–4 mm. long. Samara 3–4 cm. in diameter, usually green. Fig. 7.

UGANDA. Kigezi District: Ishasha Gorge, May 1950, *Purseglove* 3411!; Mengo District: Entebbe, Oct. 1931, *Eggeling* 26 in *F.D.* 192! & 21 km. on [Kampala–] Entebbe road, Nov. 1937, *Chandler* 2030!
KENYA. N. Kavirondo District: Kakamega Forest, Dec. 1956, *Verdcourt in E.A.H.* 11554!
TANGANYIKA. Bukoba District: near Kitwe, Oct. 1931, *Haarer* 2214! & Kabale, Sept.-Oct. 1935, *Gillman* 463! & Rubare Forest Reserve, Feb. 1958, *Procter* 833!
DISTR. U2, 4; K5; T1; southern Sudan Republic, Congo Republic and Angola west to Senegal
HAB. Rain-forest, often at edges, riparian, in forest-thickets or secondary growth; 1150–1650 m.

SYN. *Hiraea pinnata* Willd., Sp. Pl. 2: 743 (1799), *nom. illegit.* Type: as for species
        Triopteris pinnata (Willd.) Poir., Encycl. Méth. Bot. 8: 108 (1808)
        Triaspis flabellaria Juss., Malpighi. Synops., in Ann. Sci. Nat., sér. 2, 13: 268 (1840) & in Archiv. Mus. Paris 3: 507 (1843); Hook., Niger Flora: 247 (1849), *nom. illegit.* Type: as for species
        Flabellaria paniculata Cavan., var. *mollis* Engl., P.O.A. C: 232 (1895); T.T.C.L.: 296 (1949). Type: Tanganyika, Bukoba District, *Stuhlmann* (B, several syntypes †)

NOTE. Due to its wide distribution and its climbing habit the leaves of this species show a wide range of variability in shape and size, but it seems that most of the East African plants are forming a cline with a tendency towards a ± lanceolate leaf-shape. If this can be proved a constant character when more material is available, these forms may be regarded as a distinct taxon.

* Smeathman's actual specimen is in the British Museum (Natural History) (BM). Cavanilhos probably based his description on a plant grown from seeds which were taken from Smeathman's gathering. It is not certain whether there is any type-specimen kept in Paris or Montpellier.

6. CAUCANTHUS

Forsk., Fl. Aegypt.-Arab.: CXI, 91 (1775)

Diaspis Niedenzu in E.J. 14: 314 (1891) & in E. & P. Pf. III, 4 (Nachtr.): 352 (1896) & in Arb. Bot. Inst. Akad. Braunsberg 6: 15 (1915) & in Verz. Vorl. Akad. Braunsberg S.-Sem. 1924: 2 (1924)

Woody climbers or upright or semi-scandent shrubs; stems with younger parts usually ± densely appressed pubescent or sericeous. Leaves spirally arranged or opposite, with 2 glands on margin near base or eglandular; stipules very small, deciduous. Inflorescences racemose or corymbose-paniculate, axillary or terminal. Flowers actinomorphic. Sepals without glands. Petals clawed (sometimes very shortly), sometimes auriculate or hastate at base, glabrous, margins wholly or partially fimbriate. Stamens glabrous with dorsifixed anthers. Ovary densely sericeous; styles truncate, shorter than to slightly exceeding stamens. Fruit with a lateral wing completely surrounding the nut, circular or broadly elliptic; dorsal wing small, obliquely lanceolate or absent.

A genus of three species, confined to eastern Africa from the Somali Republic and Ethiopia to Mozambique and Rhodesia.

Leaves opposite; petals auriculate near the base   .   1. *C. auriculatus*
Leaves spirally arranged or rarely opposite; petals
    entire or fimbriate but never auriculate   .   .   2. *C. albidus*

1. **C. auriculatus** (*Radlk.*) *Niedenzu* in Arb. Bot. Inst. Akad. Braunsberg 6: 18 (1915) & in Vorl. Akad. Braunsberg S.-Sem. 1924: 2 (1924) & in E.P. IV. 141: 35, fig. 11/N, O (1928); T.T.C.L.: 296 (1949); E.P.A.: 403 (1956); Launert in Bol. Soc. Brot., sér. 2, 35: 48, t. 5 (1961) & in F.Z. 2: 119, t. 16 (1963). Type: Kenya, Kitui, *Hildebrandt* 2821 (M, holo.!, BM, K, iso.!)

Climber up to 5 m. in length; younger stems densely covered with short soft white sericeous hairs, older stems very finely pubescent or glabrescent. Leaf-lamina ovate-cordate, 6–12 × 4–9·5 cm., acute to shortly acuminate, membranous, pubescent or glabrescent above, grey-tomentose beneath and with 2 large glands near base (usually concealed by the indumentum); petiole 1–3 cm. long, densely sericeous with usually 2 small glands above the middle. Flowers in dense axillary and terminal corymbs, ± 1·5 cm. in diameter, evil-smelling; peduncles and rhachis sericeous; pedicels 1–1·5 cm. long, sericeous; bracts ovate; bracteoles lanceolate or linear-subulate. Sepals broadly ovate from a narrowed base, 2–2·5 mm. long, sericeous outside. Petals pale yellow, ovate, 6–7 mm. long, shortly clawed, carinate, subhastate at the base, usually reflexed. Stamens with subversatile oblong anthers 2·3–2·5 mm. long; filaments somewhat fleshy. Ovary densely sericeous; styles 2·5–3 mm. long, fairly stout, sericeous. Lateral wing of samara oblong-elliptic or oblong-obovate, 1–2 cm. in diameter, with entire margins; dorsal wing absent. Fig. 8.

UGANDA. Karamoja District: Mt. Moroto, Nakiloro R., Nov. 1963, *J. Wilson* 1582!
KENYA. Turkana escarpment, Mar. 1959, *J. Wilson* 705!; Kitui District: 80 km. S. of Kitui, 31 Jan. 1957, *Bogdan* 4380A!; Teita District: Taveta, Nov. 1937, *Dale* in *F.D.* 3854!
TANGANYIKA. Shinyanga District: Tinde Hills, 17 May 1931, *B. D. Burtt* 2384!; Mbulu District: W. slopes of Mt. Oldeani, 27 June 1945, *Greenway* 7481!; Kondoa District: scarp between Kolo and Chungai, 13 Jan. 1962, *Polhill & Paulo* 1166!
DISTR. U1; K1–4, 7; T1, 2, 5, 7; Ethiopia, Mozambique, Malawi and Rhodesia
HAB. Deciduous woodland, bushland and thicket, often riparian or in rocky places, also extending into upland dry evergreen forest and bushland; 750–1800 m.

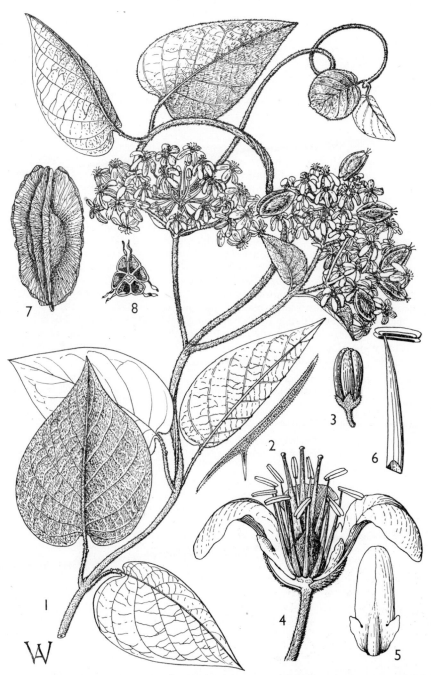

FIG. 8. *CAUCANTHUS AURICULATUS*—**1,** fertile branch, × ⅔; **2,** hair (taken from leaf), × 30; **3,** flower-bud, × 2; **4,** flower, with sepals and petals partly removed to show stamens and pistil, × 4; **5,** petal, × 3; **6,** stamen, × 6; **7,** fruit, × ⅔; **8,** fruit in transverse section, × ⅔. 1, from *Barbosa & Carvalho* 3108; 2–6, from *Anderson* 345; 7, 8, from *Dale* in *F.D.* 3854. Reproduced by permission of the Editors of " Flora Zambesiaca ".

SYN.   *Triaspis auriculata* Radlk. in Abhandl. Naturw. Ver. Bremen 8: 379 (1883);
Engl., P.O.A. A: 57 (1895), in obs.
Caucanthus argenteus Niedenzu in Bull. Herb. Boiss., sér. 2, 4: 1010 (1904).
Type: Mozambique, Boruma, *Menyhart* 964 (B, holo. †)
? C. cinereus Niedenzu in Bull. Herb. Boiss., sér. 2, 4: 1011 (1904) & in Arb. Bot.
Inst. Akad. Braunsberg 6: 18 (1915) & in E.P. IV. 141: 35 (1928). Type:
Kenya, without precise locality, *Kaiser* (B, holo. †)

2. **C. albidus** (*Niedenzu*) *Niedenzu* in E.P. IV. 141: 36, fig. 9 (1928);
E.P.A.: 402 (1956); Launert in Bol. Soc. Brot., sér. 2, 35: 47, t. 5 (1961);
K.T.S.: 259 (1961). Type: Kenya, Teita District, Ndi, *Hildebrandt* 2585 (B,
holo. †)

An upright or semi-scandent many-branched shrub; younger stems
densely covered by a silvery or whitish silky indumentum, older stems less
hairy or glabrescent. Leaf-lamina ovate, ovate-lanceolate or nearly circular,
5–30 × 4–20 mm., somewhat acute or usually finely cuspidate, sometimes ±
emarginate, usually rounded at the base, papery, densely silky on both
surfaces, sometimes glabrescent on upper surface when older, usually without
glands; petiole 1–6 mm. long, silky. Flowers in many-flowered dense
raceme-like inflorescences, whitish or cream-coloured, ± 10 mm. in diameter,
sweetly scented; peduncles densely silky; pedicels 5–10 mm. long, appressed
silky; bracts and bracteoles subulate, usually provided with 2 small glands
at the very base, deciduous. Sepals ovate, ± 1 mm. long, ± silky. Petals
ovate, 3·5–5 mm. long, somewhat carinate and hooded at the top, shortly
clawed, entire or fimbriate (usually just along one side), usually reflexed.
Anthers elliptic or oblong, 1–1·3 mm. long; filaments filiform, 2·5–3·2 mm.
long. Ovary densely hairy; styles usually 2, 1·5–2 mm. long, thick, pubescent.
Lateral wing of samara circular, 1–1·5 cm. in diameter, often with margins
crenulate; dorsal crest semi-lanceolate, ± 6 × 1·5–2 mm.

KENYA. Northern Frontier Province: Dandu, 2 May 1952, *Gillett* 13012!; Kitui
District: between Kitui turn-off and junction on Garissa–Thika road, 3 Feb. 1956,
Greenway 8855!: Teita District: near Maungu, Nov. 1937, *Dale* in *F.D.* 3757!
DISTR. **K**1, 4, 7; Somali Republic and Ethiopia
HAB. Deciduous bushland and semi-desert scrub; 180–900 m.

SYN.   *Diaspis albida* Niedenzu in E.J. 14: 314 (1891) & in E. & P. Pf. III, 4 (Nachtr.):
352 (1896) & in Arb. Bot. Inst. Akad. Braunsberg 6: 16 (1915) & in Verz.
Vorl. Akad. Braunsberg S.-Sem. 1924: 2 (1924)
D. albida Niedenzu var. *fimbripetala* Niedenzu in Verz. Vorl. Akad. Braunsberg
S.-Sem. 1924: 2 (1924). Types: Ethiopia, Harar Province, *Ellenbeck* 1037
(B, syn. †) & Kenya, Machakos/Teita District, near Tsavo, *Scott-Elliot* 6253
(BM, K, isosyn. !)
D. albida Niedenzu var. *fimbripetala* Niedenzu forma *tristyla* Niedenzu in Verz.
Vorl. Akad. Braunsberg S.-Sem. 1924: 2 (1924). Type: Somali Republic (S.),
near Kenya border, Malca Rie [Malkare], *Ellenbeck* 2146a (B, holo. †)
Caucanthus albidus (Niedenzu) Niedenzu var. *fimbripetalus* (Niedenzu) Niedenzu
forma *tristylus* (Niedenzu) Niedenzu in E.P. IV. 141: 36 (1928)
C. argenteus Chiov. in Fl. Somala 2: 41 (1932), *non* Niedenzu (1904), *nom. illegit.*
Type: Somali Republic (S.), Osboda, *Senni* 341 (FI, lecto. !)
C. chiovendae Cuf., E.P.A.: 403 (1956). Type: as for *C. argenteus* Chiov.

INDEX TO MALPIGHIACEAE

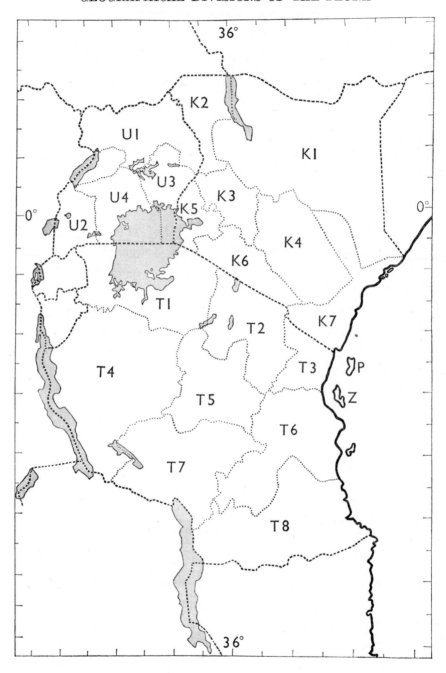